The Book (

AI for Everyday People

The Plain-English Guide to AI in Modern Life

The Book On Series

by Alex Hartman

Published by The Book On Publishing, 2025.

First Edition. May 6, 2025.

Website: https://thebookon.ca

Substack: https://thebookonpublishing.substack.com/

AI FOR EVERYDAY PEOPLE: THE PLAIN ENGLISH GUIDE TO AI IN MODERN LIFE

First Edition. May 6, 2025.

Copyright © 2025 The Book On Publishing

ISBN: 978-1-997795-85-8

Written by Alex Hartman

The Book On Series

Read This First

This is not a book designed to entertain you. It's not here to charm, to soothe, or to hold your hand. It won't dazzle you with stories, metaphors, or motivational fluff. What you're having is a tool, an instruction manual written for people who are serious about learning, executing, and thinking at a higher level.

Every book in The Book On Series is built on a single premise: clarity beats complexity. We believe that when you strip away the noise, the emotions, the marketing spin, and the cultural rituals of "self-help," what's left is raw, unembellished instruction. That's what these books offer.

They are dry by design. Not because we don't care about language or narrative, but because when you're building something that matters, you don't need more distractions. You need a clear architecture. Mental scaffolding. Direction that respects your intelligence.

Each title in this series takes on a specific domain: decision-making, clarity, strategy, leverage, uncertainty, and drills deep, not in sweeping generalizations, but in applied frameworks. These are books for builders, operators, founders, tacticians, and thinkers, people who don't just consume knowledge but operationalize it.

You'll find no chapter-long anecdotes here. No self-congratulatory memoirs. No bullet-point platitudes. Instead, what you'll get is structured insight: argument, example, application. The tone is direct. The prose is sober. The ideas are designed to be lifted out and used.

You won't be coddled, but you won't be misled either.

There's a place in the world for lyrical, emotional, story-driven books, and this isn't that place. This is a workspace. A blueprint. A conversation for people who are ready to act, not just absorb.

We respect your time and your intellect.

Welcome to The Book On Series.

Table of Contents

Dedication

For the question-askers, the skeptics, the curious minds.

This is for you.

-Alex Hartman

Preface

This book was born from conversations.

Conversations with friends who were curious about AI but didn't know where to start. I worked with coworkers who felt overwhelmed by the pace of change. With family members who'd heard the headlines but weren't sure what to believe. Again and again, I listened to the same thing: "I don't get how any of this works, but I feel like I should."

That sentence stayed with me.

Because behind it is something powerful: not fear, not cynicism, just a quiet desire to understand. And in a time when technology often moves faster than public conversation, that desire matters. It deserves a guide that doesn't assume a technical background, doesn't talk down, and doesn't sell hype.

This book is my attempt to be that guide.

It's not a deep dive into code or theory. It's a human-focused, plain-English explanation of how AI is shaping the world around us, and how you can meet it with clarity, curiosity, and confidence. It won't make you a data scientist. But it will make you literate. And in today's world, that's more than enough.

Thanks for taking this journey with me. The fact that you opened this book tells me something important:

You're paying attention. And that's where change begins.

Acknowledgments

Writing about a moving target like artificial intelligence is no small task, and I didn't do it alone.

To the researchers, journalists, educators, and ethicists whose work I've learned from, thank you for making complex topics accessible and meaningful. Your clarity gave me the confidence to do the same.

To the everyday people, friends, family, readers, who asked honest questions and admitted, "I don't get it, but I want to," you're the reason this book exists. Your curiosity reminded me who this book is really for.

To the editors and early readers who kept me honest, pushed me to simplify without oversimplifying, and reminded me to keep the heart in the work: I'm deeply grateful.

And to anyone working to make technology more human-centered, more just, and more inclusive: keep going. The future isn't built by machines; it's built by the people who choose how they're used.

Introduction

If the word "artificial intelligence" makes you picture robots, Hollywood movies, or complicated tech jargon, you're not alone.

For many people, AI feels distant, used by engineers, tech companies, or perhaps your phone's voice assistant. But here's the truth: **AI is already shaping your everyday life**, from how your favorite apps recommend content to how companies make decisions about hiring, pricing, or healthcare. Whether you realize it or not, AI is part of the world around you.

And that's precisely why this book exists.

This is not a book for programmers. It's not full of equations, code, or buzzwords. It's written for **curious, everyday people** who want to understand what AI is, how it works (in plain terms), and why it matters to your life, your job, your community, and even your future.

My goal is to give you a working understanding of AI, enough to follow conversations, spot misinformation, and make informed choices. You'll also learn how AI is being used right now, where it's going next, and what kinds of ethical and social questions it raises.

By the end of this book, you won't be an AI expert, but you will be **AI literate**, and that's something powerful in today's world.

Let's take the mystery out of AI, one chapter at a time.

- Alex Hartman

Part I: What Is AI?

When you hear the term "artificial intelligence," what comes to mind? Maybe it's a sleek, silver robot with glowing eyes and a cold mechanical voice. Or perhaps it's something a little more sinister, machines plotting to take over the world, or giant tech companies collecting your every move. Depending on what you've read or watched, AI might sound thrilling, terrifying, or just... too complicated to care about.

But the truth is both more straightforward and more surprising: **AI is already here**, and it's not just in labs or blockbuster movies. It's in your phone, your car, your streaming service. It's helping doctors detect diseases, powering translation tools, recommending what to buy, and filtering spam from your inbox. Artificial intelligence isn't the future; it's the **present**. And understanding it is no longer optional.

Yet for all its presence in our lives, AI remains a mystery to most people. Not because they're not smart enough to understand it, but because the way we talk about it is often cloaked in technical language or abstract theories. This book is about cutting through that fog, about telling the story of AI in a way that's clear, grounded, and human.

It helps to start by defining what we mean by AI. Not the science fiction version, but the real, working kind. At its core, **artificial intelligence is the ability of machines to perform tasks that typically require human intelligence**. That can mean recognizing patterns, making predictions, understanding language, or even driving a car. Some of these tasks are simple and narrow, like suggesting the next word as you type a message. Others are more complex, like helping manage air traffic or diagnosing medical conditions. But the unifying idea is this: we're teaching machines to "think," in limited ways, using data.

And that's the first important truth: **AI isn't magic. It's math.** It's based on algorithms, basically sets of instructions, and those algorithms are trained on data. The more data they have, the better they get at spotting patterns. That's how a recommendation engine knows what show you might like next. It's not reading your mind; it's just learned that people who liked what you watched also wanted something else. It's connecting dots, thousands or even millions of them.

Understanding this takes some of the mystery out of AI. But it also introduces something more interesting: a new kind of relationship between humans and machines. Not one where machines replace us (despite the headlines), but one where they increasingly **assist, augment, and shape** our decisions. This raises important questions. Who's in control? Who sets the rules? What happens when the machine gets it wrong?

Those are questions we'll explore throughout this book. But before we go there, we need a solid foundation. So, in the next chapter, we'll slow things down and get to know the basic building blocks of AI, like algorithms, machine learning, and data, in plain English. No coding. No jargon. A clear path through the buzzwords, so you can understand how these systems work.

Because once you understand the mechanics, the rest of the conversation starts to make a whole lot more sense.

If AI isn't magic, and it isn't evil robots plotting in the shadows, then what exactly is it?

To answer that, let's take a walk through what's happening behind the scenes when we say "AI." Not in tech-speak. Just in the language of regular, curious people. Because once you understand the basic moving parts, the mystery starts to dissolve, and suddenly, AI feels a lot less intimidating and a lot more human-made.

Let's start with the simplest piece: **the algorithm**.

Now, the word "algorithm" sounds like something pulled from a calculus textbook, but it's just a fancy term for a **set of rules or instructions**. Think of it like a recipe. You give it ingredients (input), follow the steps (the logic), and end up with cookies (output). In AI, the recipe might say: "If a person clicks on these types of shoes and scrolls past these kinds of shirts, recommend more shoes like those." It's just decision-making at scale, millions of tiny if-this-then-that steps, often happening in milliseconds.

But AI doesn't stop at recipes. It goes a step further: **it learns**.

This is where **machine learning** comes in, the heart of modern AI. And here's the trick: instead of programming the machine with every possible rule (which would be exhausting and probably impossible), we show it a lot of examples and let it figure out the rules on its own. That's machine learning in a nutshell: **the system finds patterns in data and adjusts itself based on what it sees.**

Imagine you're teaching a child how to recognize cats. You don't give them a strict list of rules like "four legs, fur, whiskers." Instead, you show them a bunch of pictures and say, "This is a cat." Eventually, they start recognizing cats, even ones they've never seen before, because their brain has absorbed the patterns. A machine learning system works the same way. Feed it enough examples, and it starts to make connections.

Of course, the machine doesn't "understand" in the way humans do. It doesn't know what a cat is in the emotional, tactile, whisker-twitching way we do. It knows only that the patterns of pixels in these images often occur when the word "cat" appears in the label. And if it sees those patterns again, it bets that it's probably looking at a cat. The system isn't intelligent in the conscious sense; it's **statistical**, not sentient.

This leads us to the most essential ingredient in AI: **data.**

Data is the fuel. Without data, machine learning doesn't work. That's why companies are so obsessed with gathering your data: your clicks, searches, purchases, routes to work, and even your voice commands. Every one of those actions becomes a data point that can help "train" AI systems to be more accurate, more personalized, and (from a business standpoint) more profitable.

But not all data is created equal. Data can be messy, biased, and incomplete. And that matters because if you feed an AI system insufficient data, you get bad decisions. It's the old saying: **garbage in, garbage out**. If a hiring algorithm is trained mostly on resumes from one gender or one ethnic group, it might unknowingly learn to prefer those patterns and exclude others. The system doesn't know it's being unfair. It's just following the patterns it was shown.

This is why understanding AI isn't just about knowing what it *can* do; it's about knowing **how it learns, where the data comes from, and who decides what counts as a "good" result**. These are human questions, not technical ones.

So far, we've looked at algorithms, machine learning, and data, the three building blocks of modern AI. But there's one more piece that often gets overlooked, and that's the **feedback loop**.

When an AI system makes a prediction, say, what product you'll click on, and you *do* click it, that reinforces the system's confidence. It thinks, "Aha, I was right!" (not literally, of course, but that's how it treats the outcome). Over time, the system gets better at anticipating your behavior. But it also starts shaping your behavior. If it only recommends things you already like, it might stop showing you new things altogether. That's how algorithms can create echo chambers, whether in shopping, music, or even news and politics.

So, here's the big takeaway: **AI is a system of learning rules, trained on data, that predicts or decides things based on patterns it finds.** It isn't conscious. It isn't magic. But it is powerful. And because it's designed and fed by humans, it reflects both our brilliance and our blind spots.

You don't need to know how to build an AI model to understand its impact. You need to know what questions to ask. What data is it using? Who made the model? What's it being used for? These are the kinds of questions we'll keep coming back to.

In the next chapter, we'll step back and ask: how did we even get here? Where did this idea of intelligent machines come from? How did we go from the first computers to the AI tools we carry around in our pockets?

To understand where we're going, it helps to know where we've been.

Long before artificial intelligence became a reality, it was imagined.

It began, as many powerful ideas do, in stories. Ancient myths told of mechanical beings brought to life, metal golems guarding temples, clockwork servants tending to kings, talking statues whispering the will of the gods. These early visions didn't involve code or electricity, but they carried the same spark: the idea that we might create something that could think and act like us.

Fast forward to the 20th century, and imagination met machinery.

In the 1940s and '50s, the groundwork for modern AI was laid not by computer scientists, but by mathematicians and philosophers. Chief among them was **Alan Turing**, a brilliant British thinker who asked a deceptively simple question: *Can machines think?*

Turing's answer wasn't philosophical; it was practical. He proposed a test. If a machine could hold a conversation so convincingly that a human couldn't tell whether they were speaking to a person or a computer, then for all intents and purposes, the machine could be said to "think." This became known as the **Turing Test**, and even today, it's a landmark concept in AI.

But back then, computers were still in their infancy. They were room-sized, clunky, and required punch cards or wires to operate. The idea of a machine "learning" anything seemed more like science fiction than science. Still, the dream persisted.

By the 1950s, the term **"artificial intelligence"** was officially coined at a now-legendary conference at Dartmouth College in 1956. A group of ambitious researchers gathered and made a bold claim: within a few years, they believed, machines would be able to reason, solve problems, and even learn like children. It was a thrilling prediction, but also wildly optimistic.

The decades that followed were marked by bursts of excitement and long stretches of disappointment. Early programs could play chess or solve puzzles, but only in very narrow, limited ways. They didn't generalize, and they couldn't handle the messy, unpredictable nature of real life. As a result, funding dried up, interest waned, and AI entered a period often called the **"AI winter."**

But in the background, something important was happening: computers were getting faster. Storage was getting cheaper. And most importantly, **data — the raw material AI thrives on —** was beginning to pile up in ways those early researchers could barely imagine.

Then came a turning point: the rise of the **internet** and the explosion of **digital data**.

Suddenly, every click, swipe, search, and purchase was being recorded. Social media took off. Smartphones put sensors in our pockets. This digital flood didn't just change how we lived; it created the perfect conditions for a revival of machine learning. Researchers could now train algorithms on enormous datasets. The math hadn't changed all that much, but the **scale** had. With the advent of better computers and data, AI began to accomplish things that once seemed impossible.

In 2012, a breakthrough arrived. A team of researchers at the University of Toronto trained a deep learning algorithm, a kind of layered neural network, to recognize images far more accurately than anything before it. This model, called **AlexNet**, was a game-changer. It could look at a photo and say, "That's a cat," with astonishing precision. It didn't understand cats, of course, but it had seen enough labeled images to recognize the patterns.

From there, the pace accelerated. Voice assistants like **Siri** and **Alexa** entered homes. Recommendation engines powered by AI transformed how we shop, watch, and listen. Self-driving car prototypes started navigating real roads. And AI models grew bigger and more capable, leading to tools like **ChatGPT**, which could generate human-like text in response to a question, something that would have seemed like science fiction only a decade earlier.

Today, AI isn't just a research project or a novelty. It's embedded in our everyday systems, in hiring, healthcare, education, policing, banking, and beyond. And it's evolving rapidly, often faster than the public can keep up with or fully understand.

But understanding this history matters. It reminds us that AI didn't just appear out of nowhere. It's the product of decades of effort, failure, ambition, and experimentation. It also reminds us that AI, for all its power, is still a **human creation**, shaped by the data we give it, the goals we set, and the values we embed in it.

That's what makes the next part of our journey so important. In the next chapter, we'll step out of theory and history and look at where AI is **right now** in your life. You might be surprised to find out how often you've already interacted with AI today, before breakfast, even.

Let's find out just how "everyday" this technology is.

Part II: AI in Your Everyday Life

Before you even finished reading this sentence, you may have used AI three or four times today, without realizing it.

You may have asked your phone a question, skimmed headlines curated just for you, or let your map app decide the fastest route to work. Maybe your email tucked a message into your "Promotions" tab before you saw it. Or perhaps you skipped the traffic jam this morning because your GPS app already knew what thousands of other drivers were doing.

In each of those small moments, artificial intelligence was at work. Quietly. Invisibly. Effectively.

One reason AI still feels foreign to many people is that it's no longer presented as "artificial intelligence." It doesn't announce itself. It arrives with features, tools, and conveniences, including "smart suggestions," "recommendations," and "automated responses." But behind those names are systems making predictions, scanning for patterns, and adjusting based on your behavior. They are, quite literally, learning from you, even as you go about your day, unaware.

Let's take a closer look at just how present AI already is.

Start with your **phone**. If you use voice commands, "Hey Siri" or "OK Google", you're tapping into speech recognition systems powered by AI. These tools don't just record your voice; they break down the sounds, match them to probable meanings, and try to return something useful. It's easy to forget how impressive this is: a device in your pocket is translating human speech, in real time, into machine-readable language and delivering a response. It's not perfect, but it's a remarkable leap from the dumb devices of even a decade ago.

Or take something simpler: **autocorrect** and **predictive text**. These may not seem like AI, but they are. They use models trained on vast amounts of language data to guess what word you meant, or what you'll want to type next. That moment when your phone guesses the next word in your sentence? That's a tiny AI model at work, drawing from millions of examples to predict what people like you tend to say next.

Next, consider your **entertainment habits**. If you've ever noticed that Netflix, Spotify, or YouTube seem eerily good at knowing what you'll enjoy next, that's not luck; it's machine learning. These platforms track what you've watched or listened to, compare it with the behavior of people who made similar choices, and recommend content based on those patterns. It's not about understanding you emotionally; it's about narrowing down statistical likelihoods with surprising accuracy. And when they get it right, they learn even more about you.

AI is also shaping what you **don't** see. Social media platforms, for example, use ranking algorithms to decide which posts appear at the top of your feed. These algorithms are optimized to keep you engaged, which means showing you what you're most likely to react to, whether that's a funny video, a headline, or something that stirs up strong emotions. The goal isn't necessarily to inform or inspire, it's to keep you scrolling. And AI has become disturbingly good at that.

Even the **ads** you encounter throughout the day are powered by AI, targeted, personalized, and optimized for your interests. The shoes that follow you around the internet after you browse them once? That's not a coincidence; that's a predictive model tracking your clicks, assessing your likelihood of buying, and placing bets with each ad shown.

The real point here isn't to make you paranoid, it's to make you aware. Because once you notice these invisible systems, you start to realize how deeply AI is woven into your daily life. It's not about robots taking over. It's about tools quietly making decisions on your behalf, some helpful, some annoying, and some with more profound consequences than they seem.

One of those deeper areas is **work**, how people get hired, how companies manage employees, and how decisions are made behind the scenes in business environments. Because while it's easy to see AI in our personal lives, it's also reshaping the professional world in ways that most people never see. From résumé screening to customer support chatbots to automated scheduling, AI is quietly redefining how work gets done.

That's where we're headed next.

Artificial intelligence isn't just changing the way we shop, scroll, or stream. It's quietly, and sometimes dramatically, changing the way we work.

You might not notice it right away. After all, most AI in the workplace doesn't look like a robot rolling into the office to take your job. It seems like software. It lives in the background. It filters résumés, routes emails, assigns tasks, manages calendars, and writes reports. It's not flashy. But it's everywhere.

Let's start at the beginning of the work journey: **getting hired**.

In recent years, many large companies have turned to AI-powered tools to help them manage the overwhelming number of job applications they receive. These systems don't just search for keywords anymore. Some go much further. They rank candidates based on predicted performance, analyze video interviews for tone or confidence, or use past hiring data to estimate who's most likely to succeed.

On paper, it sounds efficient. Fair, even. After all, if machines are neutral, shouldn't they be better at spotting the best candidate than a tired recruiter with unconscious biases?

The reality, though, is more complicated.

Because AI systems are trained on historical data, which often reflects human bias, they can learn to repeat and reinforce those same patterns. If a company has a history of mainly hiring one type of person for a role, the system might come to believe that's what success "looks like." It doesn't mean to discriminate, but it does. Quietly. Systematically. And without the candidate ever knowing.

This isn't just hypothetical. There have already been real-world examples where hiring algorithms were found to downgrade applications from women or candidates with nontraditional career paths, simply because they didn't fit the pattern of past hires. It's not evil intent, it's flawed input. The machine is only as fair as the data it's trained on.

And it doesn't stop at hiring.

In many offices, AI is being used to manage workflows, monitor employee activity, and even suggest when someone might be about to quit. Some systems analyze email tone to measure "sentiment" across teams. Others track keystrokes or time spent on tasks, automatically flagging when someone seems "less productive." It's workplace surveillance, but optimized.

For some employers, these tools offer insight and efficiency. Employees can feel like they are under constant, invisible oversight. The question becomes not just what these tools can do, but what they should do. How much control is too much? And who's accountable when the system makes the wrong call?

Outside the office, AI is changing entire industries.

In **finance**, algorithms make lightning-fast decisions about trades, loans, and fraud detection. In **logistics**, AI routes delivery trucks, predicts supply needs, and adjusts inventories in real time.

In **marketing**, it's used to personalize campaigns to micro-targeted audiences, sometimes down to the individual.

Even **customer service** has been transformed. You've probably chatted with a bot and not even known it. These virtual agents are getting better at handling simple tasks, answering questions, processing returns, and booking appointments. And for the companies deploying them, they're a game-changer. Less overhead. No sick days. Scalable 24/7 support.

But again, the convenience comes with trade-offs. Bots can frustrate people when they don't understand nuance. They can also hide the fact that decisions are being made by algorithms, not people, which matters when you're trying to appeal a fee, ask for an exception, or explain something out of the ordinary.

So much of this technology is built with good intentions: to speed up, to streamline, to optimize. But it all adds up to a quiet shift in how power flows. Decisions that used to be made by people, often with room for judgment or flexibility, are increasingly being made by systems designed for consistency and scale.

That doesn't mean AI is bad for work, far from it. In many cases, it's helping people do their jobs better, spotting errors, crunching numbers, and predicting trends. But it does mean we need to understand the trade-offs, and ask hard questions about transparency, fairness, and control.

As we move forward in this book, we'll keep revisiting those questions. But before we do, we need to look at another part of life where AI is showing up more and more: the places where it can affect our **health, our children's education, and even the laws that govern us.**

In the next chapter, we're going to explore how AI is being used in **healthcare, education, and government**, not in the distant future, but right now.

When algorithms start helping diagnose diseases, grade essays, or determine who gets benefits, the stakes become no longer theoretical. They're deeply human.

Something is unsettling about the idea of a machine deciding whether you get a job. But there's something even more intimate, and perhaps more worrisome, about a machine helping determine if you get medical treatment, pass a class, or receive public assistance.

And yet, that's exactly what's happening.

Across hospitals, classrooms, and government agencies, artificial intelligence is being used to sort, predict, and even influence life-altering decisions. In many cases, this technology is helping. It's catching patterns humans might miss, saving time, and easing heavy workloads. But as we've seen before, the benefits come with quiet risks, ones that are harder to see until they affect someone personally.

Let's begin with **healthcare**, a field already overflowing with data. From scans to lab results to electronic health records, modern medicine generates mountains of information. No single doctor could sift through it all, especially when trying to make quick, high-stakes decisions. That's where AI can shine.

Take medical imaging. AI systems trained on thousands, or millions, of X-rays, MRIs, and CT scans can now detect signs of disease that even experienced radiologists might overlook. Some programs are already being used to flag early signs of cancer, eye disease, or fractures. Others help triage patients in emergency rooms or monitor for post-surgical complications.

It's easy to see the potential here. Faster diagnoses. More accurate screening. Better outcomes. But these systems don't work in a vacuum; they're trained on real patient data, collected in real hospitals, often with very human imperfections. If an algorithm was mainly trained on data from one region, one ethnic group, or one type of hospital, it might perform poorly when applied to different populations. And when the algorithm gets it wrong, when it misses a cancer, or misjudges a risk, who's responsible? The doctor? The software developer? The hospital?

Doctors don't want to be replaced, and most won't be. What they want, and increasingly get, is **assistance**: decision support, not decision-making. But that line is thin, and it gets blurrier as AI becomes more advanced. As a patient, you may never know when an algorithm helped shape your care. And if something goes wrong, it's not always clear who's behind the curtain.

Now shift to **education**, where AI is appearing in subtler but still impactful ways.

You may have encountered automated essay grading systems. These tools evaluate grammar, structure, and even coherence using natural language processing. Some schools use them to help overwhelmed teachers save time on grading, while others use them in large-scale standardized testing environments. The idea is efficiency, but it raises questions: Can a machine truly judge the creativity or nuance of a student's writing? Does it reinforce a particular style over original thinking?

AI is also shaping how students learn. Adaptive learning platforms, think digital tutors, adjust the difficulty of lessons based on a student's performance. These tools can help identify gaps, provide personalized feedback, and support students who might otherwise fall behind. When used thoughtfully, they can be empowering. But they're only as fair as the models behind them. A poorly designed system might misjudge a student's ability, mislabel them, or fail to recognize cultural and linguistic differences.

Then there's the digital divide. Not every student has access to the same technology, devices, or broadband speed. AI-powered tools may seem like the future of education, but only if you're lucky enough to have access to them. Without careful planning, technology can widen gaps it was meant to bridge.

Finally, we come to **government**, perhaps the least visible and most complex area of AI deployment.

Around the world, public agencies are experimenting with algorithms to manage everything from benefit applications to child welfare investigations to criminal sentencing recommendations. In theory, these tools can make systems more efficient and consistent. But in practice, they can introduce or amplify serious problems, mainly when they operate without transparency.

One widely discussed example is the use of **predictive policing**. Some cities have trialed systems that analyze crime data to predict where crimes are likely to occur, sending more officers to those neighborhoods. But what happens when the data used to train the system is already skewed by decades of biased policing? The system doesn't fix the problem; it learns it. And the cycle continues, invisibly justified by the "objectivity" of a machine.

In another case, a state-level benefits agency used an algorithm to detect fraud and overpayments. The system flagged thousands of recipients. Many were innocent. Appeals processes were opaque, slow, and in some cases, nearly impossible to navigate. Real lives were upended by a system that wasn't malicious but wasn't accountable either.

This is the paradox of AI in public life: it promises efficiency but risks dehumanization. It's not that machines are heartless villains; it's that systems optimized for speed or scale often lack the room for empathy, judgment, or exception. And when decisions are automated, it's easy to lose track of the human story behind each case.

So, what can be done?

It starts with **transparency**. People should understand when and how AI is being used, as well as their options if they disagree with the outcome. It also requires better oversight, stronger ethical standards, and, most importantly, **informed citizens** who understand how these systems work and what questions to ask.

That's where you come in.

You don't have to be an engineer to care about how AI shapes your health, your child's education, or your access to public services. You have to pay attention, ask questions, and stay curious. That's how systems get better: when the people affected by them stay informed, speak up, and demand fairness.

In the next chapter, we'll move from how AI affects our individual lives to how it reflects, and sometimes reinforces, the biases and blind spots of the society that created it. Because as we'll see, AI isn't neutral. It learns from us. And sometimes, that's the problem.

Part III: The Human Side of AI

One of the biggest myths about artificial intelligence is that it's neutral.

People often assume that because AI is based on math, on numbers and patterns and logic, it must be fairer, more objective, and more trustworthy than human judgment. After all, a machine doesn't hold grudges. It doesn't get tired or emotional. It just does what it's told.

But here's the thing: **machines don't make themselves.** Humans build them. Humans feed them data. Humans define their goals and reward functions, and what "success" looks like. And in doing so, we quietly pass along all kinds of assumptions, some obvious, some deeply hidden, into the very heart of these systems.

That's how bias creeps in.

Let's break this down not in abstract terms, but with real-world examples.

In 2018, a major investigation uncovered that an AI system used in hospitals across the U.S. was assigning lower risk scores to Black patients compared to white patients with similar medical conditions. Why? The algorithm was using healthcare spending as a proxy for health needs, assuming that people who spent more on healthcare needed more care. But Black patients, often due to unequal access and systemic bias in the healthcare system, were spending less. So, the algorithm assumed they were healthier and deprioritized them.

No one programmed the system to discriminate. But it happened anyway.

In another case, a facial recognition system used by law enforcement was found to be significantly less accurate at identifying people with darker skin tones. Some models misidentified Black women nearly 40% of the time, compared to virtually zero errors for white male faces. Again, the problem wasn't intent; it was training data. These systems were developed using image datasets that skewed overwhelmingly toward white, male faces. The machine learned to recognize what it saw most.

These are not minor errors. They're not quirks. They are failures with real consequences, wrong arrests, delayed care, and lost opportunities.

What makes this particularly dangerous is how **invisible** algorithmic decisions can be. If a human judge, teacher, or doctor makes a biased decision, at least there's a person to question. But when a machine makes a call, especially if that machine is branded as "smart" or "advanced", we tend to trust it. Or worse, we don't even realize it's decided at all.

This is why **fairness in AI** is not just a technical problem; it's a human one. It's about whose data is used, how outcomes are measured, and who's in the room when these systems are being built.

Even well-intentioned systems can go wrong when they're trained on the past.

Imagine an AI system used to decide which job applicants should advance to the interview stage. It's trained on ten years of hiring data. But suppose that data reflects years of mainly hiring men, or mostly graduates from elite universities, or mostly people with a particular kind of résumé. In that case, the AI will learn those preferences, even if they were never spoken aloud. It will simply learn: *This is who gets hired.*

That's how discrimination gets automated: quietly, at scale, without a villain in sight.

And that's the other limit of AI: it doesn't know *why* a pattern exists. It doesn't understand fairness. It doesn't ask, "Should I be doing this?" It just optimizes. If it's told to maximize click-throughs or reduce churn or lower costs, it will do that with ruthless efficiency, even if it causes harm along the way.

So, what can we do?

First, we need to remember that **transparency matters**. Not every algorithm should be a black box. When a system is making decisions about people's lives, they deserve to know what factors went into it, what options they have to appeal, and whether there's any room for context.

Second, we need **diverse voices in the room**, not just engineers and data scientists, but ethicists, educators, community leaders, and people from all walks of life. AI isn't just a technical tool. It's a social force. And it needs social wisdom to be built responsibly.

Finally, we need to accept that AI will never be perfect. That's okay. The goal isn't perfection, it's awareness. If we understand how these systems can fail, we can design them more thoughtfully. We can push for accountability. And we can stay curious, instead of unquestioningly trusting every "smart" system that comes our way.

In the next chapter, we'll ask a different kind of trust question: not just "Is AI fair?" but **"Can we trust it at all?"** As these systems grow more powerful and more autonomous, what safeguards are in place? What risks do they pose? And how do we balance innovation with responsibility?

Let's talk about trust, transparency, and what it means to put machines in charge.

Trust is one of those things we give freely to the familiar and cautiously to the unknown. We trust the barista to make our coffee, the doctor to read our test results, and the GPS to get us where we're going. Sometimes we trust by habit. Sometimes we trust because we have no other choice.

But what happens when the system making the decision isn't a person, but a machine?

That's the question we're now facing with artificial intelligence. Not just in theory, but in practice, in everyday life, and often without realizing it. Because the reality is: **we already**

trust AI. Every time you let an app suggest your next song, or take a recommended detour on your commute, or rely on a spellchecker to fix your words, you're putting a little trust in a machine.

And for the most part, it works, which is why it's so easy to keep trusting. Until, of course, something breaks.

We've all had moments where a system fails us. The wrong address. The misinterpreted voice command. The incorrect credit card block. Most of the time, it's annoying but harmless, a glitch. But as AI becomes more capable, more complex, and more autonomous, the consequences of failure grow.

Take self-driving cars.

The idea is seductive: safer roads, fewer accidents, less human error. And there's real progress being made. These vehicles can process massive amounts of information, including speed, distance, road conditions, and pedestrians, and make split-second decisions. But they also make mistakes. They struggle in unfamiliar situations. They rely on training data that doesn't always match real-world chaos. There have already been fatal crashes involving autonomous vehicles, tragic reminders that even advanced AI can get it wrong.

So how much trust is too much?

It's not just about driving. AI is now helping pilots, diagnosing diseases, detecting fraud, and even controlling parts of our energy grid. In some military applications, AI systems are being trained to identify potential threats and recommend (or even carry out) responses. These are not science fiction scenarios. They are real systems being tested and deployed around the world.

And they raise the stakes of trust.

When we put powerful decisions in the hands of machines, particularly those involving safety, justice, or life and death, we must ask hard questions. Not only *can* this system do the job, but *how does it do the job? What happens when it fails? And who is accountable?*

One of the biggest challenges in trusting AI is that many of today's most powerful systems are what we call **"black boxes."** That means their inner workings are complex to interpret, even for the people who built them. You can feed the same system the same input twice and get two different outputs. You can ask, "Why did it make this decision?" and get no clear answer. This lack of transparency makes trust fragile, because trust without understanding is more like hope.

And this isn't just a technical issue. It's an ethical one. In situations where people are affected by a denied loan, a delayed diagnosis, or an algorithmically determined prison sentence, **the ability to question, understand, and appeal** a decision is fundamental to fairness. If we lose that, we risk surrendering too much to systems we can't see inside.

Still, there's a temptation to lean into AI. To trust it unquestioningly, especially when it seems to outperform us. After all, AI can process more data than any human, stay consistent, and avoid some of the emotional flaws that trip us up. But machines don't understand context. They don't pause to consider the ethical weight of a decision. They optimize.

That's not trust. That's delegation.

So, how do we build trust *well*?

We start with **transparency**, designing systems that are explainable, or at least auditable, so people can understand how decisions are made. We also need **accountability and** clear lines of responsibility so that when things go wrong, there's someone to answer for it. And most of all, we need **humility**: from designers, from companies, and ourselves, about what these systems can and can't do.

Because trust is earned, not assumed.

And it should never be unconditional.

In the next chapter, we'll look at how these questions of trust extend into the world of work, not just in terms of tools and tasks, but in something much bigger: the **future of jobs**. What happens when AI doesn't just assist us, but replaces certain kinds of work altogether? What jobs are at risk, and what new roles might emerge in their place?

Let's talk about automation, disruption, and what it means to be human in an increasingly machine-shaped economy.

For as long as people have invented tools, they've worried about those tools taking their place.

The printing press made scribes nervous. The tractor changed farming forever. The assembly line put artisans on notice. And now, as artificial intelligence marches deeper into the modern workplace, a familiar question rises again, this time louder, sharper, and harder to ignore:

Will AI take my job?

The honest answer is: *It depends.*

But let's be clear from the start, this is not just a replacement story. It's a story about **transformation**. Some jobs will disappear. Others will change. Entirely new kinds of work will be born. The future of employment is not a binary of humans versus machines. It's a shifting relationship between the two.

To understand that shift, it helps to look at what kinds of tasks AI is good at.

In its current form, AI excels at **narrow, repeatable, data-driven tasks**. That means jobs that involve sorting, tagging, ranking, or optimizing information are prime candidates for automation. Think data entry, basic bookkeeping, customer service scripts, or even reviewing simple contracts. These aren't always glamorous jobs, but they've been the bread and butter of many industries for decades.

AI doesn't get bored. It doesn't make typos. It can analyze thousands of cases in seconds. So yes, some of that work is going away.

But here's where nuance matters: automation tends to affect tasks, not entire jobs.

Most jobs are a mix of responsibilities, some repetitive, some creative, some interpersonal. AI might handle the scheduling and email filtering, but it still can't walk into a meeting and negotiate a deal, resolve a conflict, or mentor a struggling colleague. It can't read the room or respond to the thousand unspoken things that make up human interaction. Not yet, anyway.

So what we're likely to see is **augmentation** before elimination. People are using AI to perform parts of their jobs more efficiently or effectively. A lawyer might use AI to summarize case law. A marketer might use it to brainstorm ad copy. A teacher might use it to generate lesson plans. In each case, AI becomes a kind of cognitive partner, a tool that extends human ability, rather than replacing it outright.

That said, specific roles **will** become obsolete. History tells us this. And AI will accelerate the shift.

Take call centers. AI-powered voice assistants and chatbots are already handling millions of customer interactions. They're not perfect, but they're improving. The same goes for warehouse automation, food delivery logistics, and even content generation, areas where the cost-benefit math is increasingly tipping in favor of machines.

This shift is uncomfortable. It's disruptive. And it hits unevenly.

Workers in lower-wage, more routine jobs often face the highest risk of automation, while those in highly creative, strategic, or emotionally intelligent roles may benefit from AI-enhanced productivity. But this creates a growing pressure: **how do we reskill, retrain, and rethink education fast enough to keep up with this change?**

That question is as much societal as it is individual. Governments, companies, and communities will need to invest in lifelong learning, not just once in school, but continuously, as the nature of work evolves. Skills like adaptability, critical thinking, communication, and digital literacy will matter just as much, if not more, than knowing how to operate a specific tool.

It also raises a deeper question about **meaning**.

What is work for? Is it just about income? Or is it about identity, purpose, and contribution? If machines can do the "productive" tasks better and faster, where does that leave us?

Some believe this could usher in a future where we focus more on human-centered work, caregiving, education, the arts, and community leadership. Others see the rise of new industries entirely: AI ethics, algorithm auditing, virtual experience design, and human-AI collaboration coaching. The job titles of the next decade may not exist yet, but the seeds are already being planted.

Of course, not everyone will benefit equally. Without intentional effort, AI could widen economic divides. Companies that adopt it effectively may grow richer. Workers who can't keep pace with the changes may be left behind. The result? Greater inequality, greater anxiety, and a more profound sense of being excluded from the future.

So how do we move forward?

It starts with a shift in mindset, from fearing automation to **preparing for adaptation**. AI is not the end of work, but a significant evolution in it. The better we understand what AI can do, and what it can't, the better equipped we'll be to find our uniquely human place in the mix.

That brings us to our next question: **What is "uniquely human" anyway?**

In the next chapter, we'll explore something surprising: creativity. Because while we often think of machines as logical and humans as imaginative, the lines are starting to blur. AI is now writing stories, composing music, and generating art.

So, the question becomes: If AI can be creative, what does that mean for us?

Let's find out.

Part IV: Looking Ahead

If you had asked someone twenty years ago which human quality AI would never replicate, creativity would've topped the list.

Creativity, after all, is supposed to be the most mysterious part of us. It's the lightning bolt of inspiration, the spark in the mind's eye, the act of making something from nothing. Machines, we assumed, could follow instructions, crunch numbers, maybe even simulate thought, but **they couldn't feel**. They couldn't imagine.

And yet, here we are.

Today, AI can generate music that sounds like Bach, paint portraits that win art contests, write poems that mimic the style of Sylvia Plath, and even generate entire short stories based on a few prompts. Some of it is clumsy. Some of it is astonishing. And all of it raises a disorienting question:

Is this creativity? Or is it something else?

To answer that, we need to clarify what AI is doing when it creates.

Let's take writing, for example. A language model like the one you're interacting with now is trained on billions of words. It learns the statistical patterns of language, the likelihood of one word following another, how sentences are structured, and what tones or styles look like on the page. When it writes a story, it's not imagining a world the way a novelist does. It's assembling a plausible sequence of words based on what it has seen before. It's remixing, not inventing.

That doesn't make it useless. In fact, for many writers and artists, AI has become a kind of **creative partner**, a brainstorm buddy that never sleeps. It can help generate rough drafts, suggest ideas, or offer variations you might never have considered. It's fast, flexible, and unafraid of failure. That makes it a powerful tool.

But creativity isn't just about production. It's about **intention**.

When a human creates, there's often a reason behind the work, something we're trying to say, something we feel. Whether it's a political statement, a personal memory, or a mood we can't quite explain, art is as much about expression as it is about execution. AI doesn't have that. It has no inner life. No story it's trying to tell. No curiosity about the world.

So, while AI can **mimic** creativity, sometimes eerily well, it doesn't possess the emotional engine that drives it in humans. What it does have, though, is the ability to **augment** our creative process in unexpected ways.

In music, for instance, artists are using AI to explore new sounds, compose harmonic progressions, or generate entire arrangements they can build on. In the visual arts, designers use AI tools to test layouts, iterate styles, or even resurrect old forms with a modern twist. Some writers use AI to break through blocks, generate character dialogue, or challenge themselves with plot turns they wouldn't have thought of on their own.

This doesn't mean AI replaces the artist; it **reframes** the artist's role. Less solitary genius, more creative director. Less blank canvas, more co-author. The creator becomes someone who guides, selects, curates, and reshapes what the machine offers.

Of course, this shift comes with tension.

In 2022, a digital artwork partially generated by AI won first prize at a fine art competition in Colorado. The backlash was swift. Some accused the artist of cheating. Others mourned the idea that a machine could be considered an "artist." And underneath all of that was a deeper worry: **Are we losing something sacred?**

That's the heart of the conversation.

When we see machines doing things we once considered uniquely human, such as writing, painting, and composing, it forces us to ask: what makes our creativity meaningful? Is it the product? The process? The person behind it?

For some, the answer is clear: human creativity is irreplaceable because it's grounded in life experience. AI can copy, but it can't feel heartbreak. It can't fall in love. It can't struggle through doubt, wrestle with memory, or find catharsis in creation. And that's what gives human art its soul.

Others see it differently. They argue that AI is expanding the boundaries of what creativity can look like, that art has always been about tools, and this is just the next evolution. The camera didn't kill painting. Synthesizers didn't destroy music. Maybe AI won't kill creativity either; it will just change its shape.

Wherever you land on that spectrum, one thing is clear: **the conversation is just beginning**.

As AI becomes more capable, the line between "real" and "machine-made" will blur. Some people will embrace it. Others will resist. But all of us will need to grapple with what it means to create, express, and connect in a world where machines can imitate the appearance, if not the spirit, of imagination.

In the next chapter, we'll take this conversation even further, into the realm of rules, rights, and responsibilities. As AI spreads into art, health, law, and labor, **governments and institutions are scrambling to catch up**.

Let's discuss ethics, regulation, and the challenge of setting boundaries for something we're still learning to understand.

For much of its life, artificial intelligence was an experiment, something confined to labs, university papers, and the imaginations of science fiction writers. But now, AI is no longer a side project. It's embedded in our phones, our homes, our schools, our courts, our hospitals, our jobs. It's making decisions, sometimes big ones, and it's doing it at a scale and speed that few people ever imagined.

And the truth is, we're not quite ready for it.

AI has moved faster than our laws, faster than our institutions, and in some cases, faster than our ability to understand it. So we find ourselves in a moment of catch-up: trying to set guardrails for a technology that's already out on the road, picking up speed.

It's not that nobody's trying. There's a growing movement among lawmakers, ethicists, researchers, and concerned citizens to figure out how to govern AI wisely. But it's not easy. Because, unlike a new car or a pharmaceutical drug, AI doesn't fit neatly into existing regulatory boxes.

It doesn't live in one place. It can't be inspected with a single checklist. And because many systems are developed by private companies, their inner workings are often proprietary, protected as trade secrets. That makes it hard for regulators, or even the public, to see what's going on under the hood.

So where do we begin?

For many, the answer starts with **ethics**.

Before we write laws, we need to ask: What *should* these systems be allowed to do? What values should guide their design and use? And who gets to decide?

One of the core ideas in AI ethics is **accountability**. If a system denies you a loan or makes an error in a medical diagnosis, who is responsible? The developer who built the algorithm? The company that deployed it? The machine itself? Right now, the answers vary wildly depending on where you are and who's involved, and often, there's no clear answer at all.

Another ethical principle is **transparency**. People should have the right to know when they're interacting with AI. They should understand how decisions are being made about them, and what data is being used. This is especially critical in high-stakes areas like healthcare, policing, or employment. A "black box" model that no one can explain doesn't just feel unfair; it's **unaccountable by design**.

And then there's **fairness**, the idea that AI systems shouldn't discriminate or reinforce harmful biases. We've already seen how this can go wrong in previous chapters. What's needed is not just technical fixes, but broader conversations about whose values are built into the system, whose experiences are considered, and who has a voice in the design process.

So how are governments responding?

The answer: unevenly.

Some countries are charging ahead. The European Union, for instance, has proposed sweeping AI legislation that would classify systems by risk level, banning some outright (like social scoring systems) and strictly regulating others, such as facial recognition or hiring algorithms. The EU's approach focuses heavily on human rights, transparency, and public accountability.

Other countries, including the United States, are taking a more fragmented path, issuing guidelines and recommendations, but so far avoiding broad federal laws. Some cities and states have passed their own rules, especially around surveillance and policing. But the landscape is patchy, and enforcement often lags behind the pace of development.

Meanwhile, **private companies** are setting their own rules, drafting AI ethics statements, hiring internal oversight teams, and sometimes halting products that pose public risk. While this is better than nothing, relying on companies to self-regulate is risky. Their bottom line is still profit, and ethical choices can be costly. Without external pressure, whether from laws, public opinion, or investor demands, ethical promises may stay on the shelf.

So, what comes next?

One thing is sure: **regulation is coming.** The question is how well it will be designed, and who will be included in the process, because AI doesn't just affect tech companies or scientists. It affects all of us. And shaping the rules around it shouldn't be left to a small circle of experts behind closed doors.

This is why **AI literacy** matters. The more we understand what these systems do, how they work, and where they're used, the better we can engage in these conversations. That's what this book has been building toward, not technical mastery, but confident, informed awareness.

Because the future of AI isn't something that will be handed to us, it's something we'll shape together.

So in our final chapter, we'll bring everything together. We'll revisit the key ideas, offer some practical advice, and ask the most important question of all:

What kind of relationship do you want to have with AI, and how do you stay human in the process?

Let's turn the page and get into fundamentals and real world scenarios..

Part V: What Would You Do?

Understanding ethics isn't just about knowing what's right or wrong. It's about making decisions in complex, often gray areas, where trade-offs matter and consequences ripple outward in ways we can't always predict.

When it comes to artificial intelligence, these moments are happening already, in companies, governments, classrooms, hospitals, and homes. They're not always labeled as "ethical dilemmas," but that's precisely what they are.

Let's walk through a few real-world-inspired scenarios.

Dilemma #1: The AI Hiring Tool

You work in HR at a growing company. The executive team is excited to adopt an AI-powered résumé screening tool that promises to speed up hiring and reduce bias.

After three months, you discover something troubling: the system seems to prefer candidates from certain universities and consistently ranks applicants with nontraditional backgrounds lower.

The vendor claims the system is "data-driven and fair." But you're seeing patterns that suggest otherwise.

What would you do?

> ➢ Do you pause use of the system and raise concerns to leadership, knowing it might delay hiring?

> ➢ Do you dig deeper into the algorithm's logic, even if it means pushing past confidentiality clauses?

> ➢ Or do you keep using it, assuming it's still better than relying on gut instinct?

Dilemma #2: The Personalized Learning App

You're a teacher in a public middle school. A new adaptive learning app promises to customize math lessons to each student's ability level. At first, it seems to help struggling students catch up.

But over time, you notice something strange: some students are rarely shown more rigorous material. When you investigate, the app's algorithm had "decided" they weren't ready, based on just a few low test scores.

Now those students are behind. The app says it's just following the data.

What would you do?

➢ Do you override the app and manually assign harder lessons, even if the system pushes back?

➢ Do you remove the app entirely, losing the benefits it offered others?

➢ Or do you work with developers to push for more transparency and control?

Dilemma #3: The Predictive Health Model

You're part of a public health team planning how to allocate resources for chronic disease prevention in a large city. An AI model predicts which neighborhoods are at the highest risk for future outbreaks of diabetes and heart disease. However, the data it relies on includes income, zip codes, and other demographic markers, making it difficult to determine whether it's revealing health risks or just reflecting **systemic inequality**.

Sending extra resources to these areas might help, but could also unintentionally stigmatize those communities.

What would you do?

> ➢ Do you follow the model's recommendations and target resources accordingly?
>
> ➢ Do you challenge the model and demand more equity-focused inputs?
>
> ➢ Or do you strike a balance, trusting the tool, but involving community leaders in final decisions?

Dilemma #4: The Surveillance Software

You work in local government. A tech vendor offers a new tool that uses facial recognition to monitor public spaces for known threats and criminal activity. It's being marketed as a way to improve safety.

But civil rights groups raise concerns: the system is less accurate for people of color and could be used to track lawful protests or target marginalized communities.

What would you do?

➢ Do you support deploying the tool, hoping the benefits outweigh the risks?

➢ Do you push back and demand a full audit of the system before approval?

➢ Or do you insist on pausing all use of facial recognition in public spaces?

The Point Is Not to Have the "Right" Answer

These aren't hypothetical for long. In some places, they're already happening. And the people making decisions often don't have deep technical expertise. They're educators, parents, community leaders, small business owners, and **everyday people like you**.

That's why ethics matters. It gives us a framework to ask not just *whether* we can use a system, but *whether we should.*

Because the future of AI isn't only being written in labs and boardrooms. It's being shaped, right now, by the decisions each of us makes, quietly, daily, sometimes under pressure.

So, the question is:

What kind of future are you helping build?

By now, you've seen that artificial intelligence isn't just one thing. It's not a singular invention or a looming robot overlord. It's a growing family of systems, some helpful, some flawed, some astonishing, that are increasingly shaping how we live, work, learn, and relate to each other.

We've covered a lot of ground: what AI is, how it works, where it shows up, and why it matters. You've seen how it's quietly woven into your phone, your job, your news feed, how it influences healthcare, education, government, and the arts. You've seen its promise and its pitfalls.

But what now? What do you do with this knowledge?

The answer is more straightforward than you might think; **you become AI-literate.**

AI literacy isn't about learning to code. It's not about memorizing terminology or building algorithms. It's about understanding how these systems work *well enough* to ask good questions, spot red flags, and make informed decisions in a world that they increasingly shape.

It means recognizing when you're interacting with AI, even if it's not apparent. It means asking:

> ➤ Where did this recommendation come from?
> ➤ What data is this decision based on?
> ➤ Who benefits from this system, and who might be left out?

It means **staying curious**, instead of intimidated.

And most of all, it means remembering that AI is not some alien force descending on us, it's something **we** built. Which means we have a say in how it's used. How it's designed. What it serves. What it respects.

This is where the human part comes in.

Because, as smart as machines get, they still don't understand context, compassion, justice, or meaning the way we do. They can analyze patterns, but they don't know what it's like to live with uncertainty. They don't love, or grieve, or hope. That's our territory.

So, while AI may reshape many of the tasks we perform, it can't replace what it means to be human.

It's up to us to protect that space, to design systems that reflect our values, not just our efficiencies. To build technology that helps people, rather than just managing them. And to hold space for the kinds of thinking, creating, and connecting that no machine can truly replicate.

You don't need to be an expert to be part of this future. You need to be aware, engaged, and ready to keep learning.

That's what becoming AI-literate is all about.

It usually starts with something small.

Perhaps you'll receive a smart speaker as a gift. Or you decide to try a video doorbell to feel a little safer. Or maybe you buy a new thermostat that promises to learn your habits and lower your energy bill.

At first, it feels like magic.

You dim the lights by voice. You ask your speaker to play your favorite song, and it does. You turn up the heat on your phone on a cold night. And just like that, the line between home and technology begins to blur.

Welcome to life with **AI at home**, a world where your appliances listen, learn, and adapt. Where your routines become data. And where comfort, convenience, and control come with quiet, often invisible trade-offs.

Let's take a closer look at what's happening when your home gets smart.

The Rise of the Smart Home

In the past decade, smart home technology has exploded. Voice assistants like Amazon's Alexa, Google Assistant, and Apple's Siri have become household names. Smart thermostats like Nest or Ecobee learn when you're home and adjust accordingly. Lights, locks, fridges, cameras, even coffee makers, can now be connected, controlled remotely, and in many cases, controlled by AI.

At first, these systems just followed instructions. "Turn off the lights." "Set the alarm." But over time, they've begun to anticipate. That's where AI enters the picture.

When your smart speaker starts suggesting new music based on your habits, that's AI. When your thermostat lowers itself after noticing a pattern in your absence, that's AI. When your doorbell camera flags "unusual activity" at 2:00 a.m., it's using machine learning to decide what's normal and what's not.

It's subtle. And that's part of its power.

Learning Your Life

The brilliance of home AI is how quietly it adapts. It doesn't ask much of you. It observes. It waits. It builds a model of your routine, then acts on it.

Over time, it might learn that you usually go to bed at 11:00 p.m., so the system dims the lights at 10:50. It might notice that you crank up the AC on Saturdays and start doing it for you. These are small gestures. But they add up to something that feels seamless, like your home "gets" you.

This is the dream of AI at home: **personalization without effort**.

But behind that dream is something worth understanding. For an AI to "get" you, it has to know you. And to know you, it has to collect information, often more than you realize.

The Cost of Convenience

Here's where things get complicated.

That smart speaker? It's always listening for its "wake word." In most cases, companies claim it only begins recording after you say "Hey Alexa" or "OK Google." But it still has to listen *enough* to know when you've said it. And sometimes, it records by mistake.

Those smart cameras? They upload videos to the cloud. Those smart fridges? They may track what's in your kitchen to help with grocery suggestions. Your devices are learning about your schedule, your habits, your preferences, even the sound of your voice.

And all of that data goes somewhere.

Sometimes it's used to make the product work better. Sometimes it's used to sell you more products. And in some cases, it's shared with advertisers, with third-party developers, and, in rare but real instances, with law enforcement.

Most people don't read the fine print. They click "agree" and move on. But in doing so, they're opening the door, not just to convenience, but to **surveillance**.

Privacy in the Age of Smart Everything

Privacy in a smart home isn't just about what the system sees. It's about **who controls the system**, and what happens to your data afterward.

Can you delete your recordings? Can you stop the device from sharing data with others? Who has access to your camera feeds, and for how long?

The answers vary, by company, by country, by product. And they're often buried in terms of service documents written in legalese. That means the burden falls on the user to understand what's happening behind the scenes, and frankly, most people don't have the time or technical knowledge to do that.

So, what do you do?

You stay curious. You stay informed. And you ask questions.

Do I need this device connected to the internet?

What data is it collecting, and why?

What happens if it gets hacked?

Who benefits most from this "smart" feature, me or the company selling it?

These aren't paranoid questions. They're practical ones.

Balance, Not Fear

It's easy to swing toward extremes when talking about smart homes, either a total embrace or a total rejection. But the truth, as usual, lies in the middle.

AI at home can be **genuinely helpful**. It can support people with disabilities, reduce energy waste, protect against break-ins, and streamline the chaos of modern life. But it needs to be used with intention. The more we understand what these tools are doing, the more we can use them *on our terms*, not theirs.

So, if you're building a smart home or just starting with a smart speaker, ask yourself:

Am I still in control? Or is the system deciding too much for me?

Because the future of AI at home doesn't just depend on better technology, it depends on better awareness. On knowing the line between comfort and compromise, and choosing which side of that line you want to live on.

It usually starts with a simple question.

"Is Siri a robot?"

Or maybe: "Did the computer write that story?"

Or the one that stops you cold: "Is a machine going to take my job one day?"

Kids are asking these questions earlier and more often. Sometimes they come from curiosity. Sometimes they come from what they overheard on the news. And sometimes, they come from a deep place of anxiety, even if they can't fully articulate it.

As adults, it's tempting to brush these questions off. To change the subject. To say, *"Don't worry about that, you'll learn it when you're older."*

But here's the truth:

If AI is shaping the world they're growing up in, we owe it to them to help make sense of it.

Not with technical jargon. Not with hype or fear. But with calm, honest conversations, rooted in facts, guided by empathy, and adapted to their age and understanding.

Start Simple: What AI Is

The first step is helping kids understand that AI isn't magic; it's machines doing tasks that people used to do, often by learning from patterns in data.

You might say something like:

"AI is like a fast student that learns by seeing lots of examples. If you show it a thousand pictures of cats, it can start guessing which ones are cats in new photos, even if it doesn't know what a cat is."

You don't need to explain neural networks or supervised learning. Instead, **ground it in something they already know.** If they use YouTube, you can explain that the recommendations they see come from an AI trying to guess what they'll want next. If they talk to Alexa or Google, you can explain that the voice assistant is following patterns to try to help.

This approach helps demystify the technology. It puts the power back in human hands.

Encourage Questions

Kids are natural philosophers. They want to know "why," "how," and "what if." And AI is full of questions like that:

- ➢ Can a machine have feelings?
- ➢ Can it be your friend?
- ➢ Is it spying on us?
- ➢ Can it lie?

The best response isn't to shut these questions down. It's to **explore them together**. You don't have to be an expert. Saying "That's a great question, I don't know either, but let's find out together," might be one of the most empowering things you can model.

When kids see that even adults are learning, they stop thinking of AI as a mysterious, all-knowing thing and start seeing it as something **people build, people shape, and people are still figuring out**.

Talk About the Limits

One of the most important lessons you can teach a child about AI is this:

"Just because a computer says something doesn't mean it's always right."

AI makes mistakes. It doesn't understand feelings or fairness. It might give a wrong answer, miss important context, or reflect the bias of the data it was trained on.

Explain that a computer program is **only as innovative or fair as the information it learns from**, and that humans are the ones responsible for checking its work. This can be a great way to reinforce values like critical thinking, responsibility, and the importance of asking questions.

Balance Wonder with Wisdom

AI can do some pretty amazing things. It can draw pictures, answer questions, write stories, translate languages, and even help kids with homework. It's okay to be excited about that. You don't need to crush their curiosity with warnings.

But it's also okay to share the big picture.

You can say:

"AI can be helpful, but we have to be careful how we use it. It's like a powerful tool, kind of like fire. It can light your house... or burn it down. That's why we must use it wisely."

This kind of language helps kids see AI as something they can explore responsibly, not something to fear or unquestioningly trust.

Set Boundaries and Expectations

If you're a parent or educator, you're also a guide for how kids interact with technology. That means helping them understand not just **what AI is**, but **how to use it wisely**.

You might talk about:

> ➤ Why it's important not to give personal information to a chatbot.

> ➤ Why copying homework answers from an AI app isn't real learning.

> ➤ Why taking breaks from screens matters, even when the AI is fun.

These conversations don't need to be lectures. They can be part of dinner table chats, classroom projects, or everyday observations. What matters most is that you **normalize talking about tech as something we shape, not something that shapes us.**

Let Them Lead

Finally, remember that today's kids are tomorrow's creators. Some of them will be the ones designing the next generation of AI tools. Let them play, experiment, ask "what if?" Give them safe, ethical spaces to explore how this technology works.

You might be surprised what they come up with, and how thoughtfully they respond.

Because while adults often approach AI with skepticism or worry, kids are more likely to ask:

"How can we make this better?"

And that's the question we should all be asking.

In the following (and final) chapter, we'll bring all of these ideas together. We'll revisit the big themes, offer a few parting thoughts, and remind ourselves that AI literacy isn't about mastering a machine, it's about **understanding the world we're helping build**.

The truth about the future is that no one can see it clearly, not even the smartest AI.

But we can see the shape of things. The outlines. The momentum. We can listen closely to what's happening now, pay attention to the trends, and imagine what's likely to come if we stay on our current path.

So, let's look ahead, not to make bold predictions, but to prepare.

More AI, Everywhere

Over the next decade, artificial intelligence will become more embedded in daily life and more invisible.

It won't just be the voice assistant in your kitchen or the chatbot on a website. AI will quietly shape traffic patterns, manage hospital supplies, adjust your streaming service in real time, and personalize everything from shopping to education. You might not even realize it's happening.

And that's precisely the point: AI will move further into the background. Systems will get better at seeming seamless, faster, smoother, quieter.

The question we'll have to keep asking is: "Do I still know what's going on under the surface?"

Because the more invisible AI becomes, the more critical it will be to stay aware of who's designing the systems, what goals they're optimizing for, and how those choices affect everyday people.

Fewer Jobs Replaced; More Jobs Changed

One of the most persistent fears about AI is that it's coming for our jobs. But the next 10 years may not bring mass unemployment; it may bring **mass adaptation**.

We'll likely see more **hybrid roles**, where AI handles the repetitive or analytical work and humans focus on empathy, strategy, and creativity. Some jobs will disappear. Others will be born. Most will change in ways we haven't fully mapped yet.

What will matter most isn't just what you *know*, but how quickly you can *learn*. Adaptability, curiosity, and communication will become some of the most valuable "skills" around.

And lifelong learning? That won't just be a slogan. It'll be a way of life.

AI in the Hands of the Public

Today, the most powerful AI tools are owned by big companies. But in the next decade, we'll see more efforts to **decentralize and democratize** AI.

Open-source models. Local versions of chatbots. Community-developed tools built with transparency in mind. These alternatives won't replace the tech giants overnight, but they'll offer something we badly need: **choice**.

People will begin asking not just what AI *can* do, but who gets to decide what it *should* do. The public conversation will mature, from wonder and worry to design and demand.

And that's a good thing.

New Risks, New Responsibilities

With more power comes more pressure.

The next decade will bring more challenging conversations about ethics, surveillance, misinformation, and accountability. Deepfakes will become harder to spot. AI-generated content will flood the internet, and not all of it will be harmless. Elections, economies, and even wars may be influenced, directly or indirectly, by algorithmic decisions.

Governments will try to regulate. Some will do it thoughtfully. Others will fumble. Companies will claim to self-police. Some will mean it. Others won't.

The challenge for the rest of us is to **pay attention without getting paralyzed**. To stay informed, stay involved, and keep insisting on transparency, fairness, and control.

The goal isn't perfect. It's accountable.

A Human Future, Still

Amid all the change, one thing won't go away: the need for human judgment, connection, and care.

Even the most intelligent machine won't raise a child, comfort a grieving friend, teach a struggling student, or build trust in a divided room. The things that make life meaningful can't be optimized or outsourced.

As we look ahead, the real question isn't, "What will AI do?" It's:

"What will *we* do with this tool, this moment, this power?"

The next ten years will be full of invention. But what will define them is intention.

Let's make sure we bring wisdom with us into the future we're building, because machines will never be wiser than the people who train them.

And the world doesn't need more intelligence.

It needs more humane intelligence.

ALEX HARTMAN

Case Studies

Predicting Wildfires with AI

On a dry summer morning in California, a spark from a power line ignites a brush fire. Within minutes, flames are racing across the hillside. For decades, communities have lived with this fear, waiting, watching, reacting. But now, they're getting help from an unlikely ally: artificial intelligence.

Researchers and emergency response teams are using AI systems to **predict wildfire risk**, not just based on weather forecasts, but on a blend of data points too complex for any human to process alone. These systems take in satellite imagery, soil moisture levels, wind patterns, past fire behavior, and even the health of vegetation, then generate detailed maps of areas most at risk in the coming days.

In some counties, this information now informs where firefighters are stationed, where evacuations are planned, and how resources are distributed before a fire even starts.

It's not perfect. AI can't stop lightning, sabotage, or climate change. But it gives communities something rare: **a head start**. And when minutes can mean the difference between containment and catastrophe, that matters.

This is AI not as a product, but as a **public good**, a tool used quietly, in the background, to help save lives and land.

Adaptive Learning in a Chicago School

Ms. Ramirez stands at the front of her eighth-grade math class in Chicago, watching her students work through equations on their laptops. But something is different this year. Her students aren't all working on the same problems. Each one is tackling material tailored to their level, pushed just enough to challenge them, but not so much that they give up.

What makes this possible is the **adaptive learning platform** her school adopted, a form of AI that tracks each student's progress in real time. The system notices when someone struggles with fractions, slows down, and offers extra practice. If another student races through algebra, it moves them forward. The software doesn't replace the teacher; it gives her a kind of digital assistant, helping her see patterns across the classroom that might otherwise go unnoticed.

One of her students, Jamal, was flagged last year for consistently underperforming on standardized tests. But the adaptive system revealed something else: he just needed more time. With extra support, Jamal began to close the gap. Not because he got smarter, but because the system got **smarter about how he learned**.

Of course, Ms. Ramirez still plays the most significant role. She encourages, explains, and intervenes. But now, she has tools that let her meet students where they are, not just where the curriculum says they should be.

Diagnosing Diabetic Retinopathy in India

In rural India, access to eye care is a luxury. Millions of people live hours from the nearest specialist, and preventable conditions like diabetic retinopathy, the kind that can lead to blindness, often go undetected until it's too late.

But in dozens of clinics, a quiet revolution is happening.

Using a smartphone-based AI tool, healthcare workers can now take a picture of a patient's eye and receive an instant assessment. The system, trained on hundreds of thousands of retinal scans, can detect signs of disease with remarkable accuracy, sometimes **more accurately than general physicians**.

This doesn't mean doctors are no longer needed. But in places where doctors are stretched thin, the AI serves as a **front-line triage tool**, flagging high-risk patients so they can be referred for urgent care.

The result? More early diagnoses, fewer missed cases. And more people are keeping their sight.

It's a reminder that AI doesn't have to be flashy. Sometimes, it just needs to be **available, reliable, and quietly effective**, especially in places where healthcare resources are scarce.

Reducing Bias in the Courtroom, Or Not?

In parts of the U.S., judges are now using risk assessment algorithms to help decide whether someone awaiting trial should be granted bail. The idea is simple: take past data about similar defendants, age, criminal history, past court appearances, and predict the likelihood that a person will reoffend or skip their trial.

But what sounds like neutral math has sparked intense debate.

In one high-profile case, two defendants were arrested for similar charges. One was flagged as low risk, the other as high. But years later, reporters revealed that the algorithm had made these predictions in direct opposition **to what happened**. The higher-risk defendant never reoffended. The lower-risk individual committed a violent crime.

Digging deeper, researchers discovered that the algorithm's predictions were often **skewed by race and socioeconomic factors**, not because the model was explicitly racist, but because the historical data it learned from reflected decades of biased policing and sentencing.

It's a chilling reminder that even the most powerful tools can **inherit the flaws of the systems that built them**. And when they're used in matters of justice, those flaws aren't just academic, they're deeply personal.

Watching Earth from Above, AI in Climate Monitoring

From a satellite orbiting 500 miles above the Earth, a stream of high-resolution images beams down every day. To the human eye, it's a blur of forests, cities, oceans, and clouds. But to an AI system developed by climate researchers, it's something more: **a**

living, changing planet, full of warning signs and quiet transformations.

These AI tools are trained to spot subtle shifts in land use, deforestation, glacier movement, and even methane leaks from oil fields. One model flagged an increase in illegal logging in the Amazon long before local authorities noticed. Another helped estimate crop failures in drought-stricken regions, enabling earlier aid distribution.

The scale is breathtaking: AI can analyze **millions of square kilometers of satellite data in hours**, finding patterns that humans would miss. It's becoming a global early-warning system for environmental risks.

This isn't just about data. It's about survival. And in a world where every day matters in the fight against climate change, having a system that never sleeps, never slows, and never stops watching... might make all the difference.

Giving Voice to the Voiceless, AI and Accessibility

For years, 19-year-old Sarah, who lives with severe cerebral palsy, relied on a slow, clunky communication board to express herself. Conversations were slow. Spontaneity was nearly impossible. One day, her parents introduced her to a new AI-powered app that changed everything.

Using a blend of predictive text, voice synthesis, and eye-tracking, the app allowed Sarah to form complete sentences just by moving her eyes. The AI system learned her patterns, how she phrased things, which words she used most, and adapted over time. What used to take five minutes to say could now be said in seconds.

Her world opened up.

She started texting friends, asking questions in class, and telling jokes. Her personality, long buried behind a wall of effort, came shining through.

This is AI at its most humane, not replacing human interaction, but **restoring it**, for someone who had been unintentionally left out.

Fighting Fraud in Real Time

When Maya got a late-night text from her bank asking if she'd just purchased $1,800 worth of electronics in a city she'd never visited, she panicked. But within seconds, the transaction had been blocked.

Behind the scenes, an AI fraud detection system had flagged the transaction as unusual. Based on her spending habits, location, amount, and timing, it didn't match her profile. The AI wasn't just following a static rule like "over $1,000 = bad." It was constantly learning her patterns and updating its understanding.

Banks around the world now rely on these systems to detect credit card fraud, phishing attempts, and identity theft. They process **thousands of transactions per second**, flagging anomalies with increasing accuracy. And when they get it right, they don't just save money, they **protect trust**.

Not every system is perfect. Sometimes the AI overcorrects. But the net effect is apparent: catching crime at the speed of data is something no team of humans could do alone.

Reader Questions

A Conversation About AI

Throughout writing this book and talking with friends, readers, and workshop audiences, I've heard a lot of great questions. Some were practical. Some were philosophical. Some came from curiosity, others from concern.

Here are some of the most common questions people ask when they're trying to wrap their heads around AI. My answers are as straightforward as possible, with no jargon, no hype, just the real talk you deserve.

Q: Is AI really "intelligent"? Or is that just marketing?

It depends on how you define "intelligent."

AI excels at specific tasks, including recognizing patterns, processing massive amounts of data, and generating text or images based on examples. But it doesn't *understand* what it's doing. It doesn't think, feel, or reflect. So in that sense, it's not intelligent like a person; it's more like a very fast, very confident guesser.

Think of it this way: AI is not your new best friend. It's a glorified spreadsheet with imagination.

Q: Should I be worried about AI taking over my job?

Worry? Maybe a little. But more importantly, stay *prepared*.

AI is already changing how we work, automating some tasks, assisting with others, and sometimes making specific roles less necessary. But most jobs won't disappear overnight. They'll evolve.

What you can do is stay flexible, keep learning, and focus on the skills that machines aren't good at: creativity, empathy, leadership, judgment. AI can write an email. It can't lead a team or care about one.

Q: Can AI be creative? Like, creative?

It can produce things that look creative, such as songs, stories, and images, but it's remixing, not inventing.

AI creates based on patterns in what it's already seen. It doesn't know why a poem makes someone cry. It doesn't feel the joy of making something new. So yes, AI can mimic creativity. But it doesn't *mean* anything by it. That's still our domain.

In short, AI can generate. Only humans can create with meaning.

Q: If AI makes decisions based on data, isn't that fairer than human judgment?

That's the theory. The reality is trickier.

AI is only as fair as the data it's trained on, and human data is full of human flaws. If the past was biased, the system learns bias. If the data excludes certain groups, the system may inadvertently ignore or harm them.

So no, AI isn't automatically fair. It can lock in unfairness even faster, because it does it at scale, invisibly, and without asking hard questions.

That's why we need people to stay informed.

Q: What's the most significant risk with AI right now?

It's not evil robots or rogue superintelligence. It's **misuse by humans**, whether that's surveillance without consent, automated systems that discriminate, or tools used to manipulate public opinion.

The most significant risk is assuming that just because something works, it's okay to use it.

We're still learning what AI is good for. In the meantime, we need to lead with ethics, not just efficiency.

Q: I don't work in tech. Why should I care about this?

Because AI is shaping things that *do* affect you: what jobs are available, how healthcare is delivered, what content you see, and what decisions are made about you in schools, banks, and government offices.

You don't need to be a programmer to be part of the conversation. You need to be aware. AI literacy isn't about code, it's about **citizenship in a tech-shaped world**.

Q: How do I spot AI-generated content online?

It's getting harder every day. Some giveaways include weird phrasing, repetition, generic statements, or oddly "perfect" grammar. But newer models are improving fast.

Look for context clues: Is the source reputable? Is there a real author? Does the content feel oddly hollow, even if it's technically correct?

Trust your instincts, and when in doubt, double-check with other sources. A little skepticism is healthy.

Q: What's one thing I can do right now to feel more in control of AI in my life?

Pay attention.

Start noticing where AI shows up, in your apps, your searches, your recommendations. Ask questions. Get curious. When you understand how something works, it stops being magic and starts being manageable.

You don't need to master everything. You need to stay *awake*.

Further Reading & Resources

If this book sparked your curiosity and you want to dig deeper, here are some books, podcasts, and tools that balance accessibility with insight.

Books

➤ *You Look Like a Thing and I Love You* by Janelle Shane
(A witty, beginner-friendly look at AI gone right and wrong.)
➤ *Weapons of Math Destruction* by Cathy O'Neil
(A powerful exploration of how algorithms can harm when left unchecked.)
➤ *Artificial Unintelligence* by Meredith Broussard
(Why we shouldn't expect AI to solve everything, and what we should focus on instead.)

Podcasts

➤ **Hard Fork** (New York Times tech journalists break down major AI trends)

➤ **AI Alignment Podcast** (For deeper ethical discussions and interviews with researchers)

➤ **The Ezra Klein Show**, select episodes on technology and ethics

AI Tools to Explore

➢ **ChatGPT (OpenAI)**, Try interacting with an AI language model in real time

➢ **DALL·E or Midjourney**, generate images from text prompts

➢ **Teachable Machine (by Google)**, Create simple machine learning models with no coding

Online Courses & Guides

➤ *Elements of AI* (Free, beginner-level course developed in Finland)

➤ *AI for Everyone* by Andrew Ng (Coursera)

Chapter Reflections

Chapter 1: Welcome to the Age of AI

You've seen that AI is already here, not in the distant future, but woven into your daily routines.

What's one example of AI you now realize you've interacted with recently?

Pause and consider how that tool made your life easier... or perhaps more complicated. How does it feel knowing that something you once considered high-tech is already part of your typical day?

Chapter 2: AI in Plain English

Now that you understand the basics, algorithms, machine learning, and data, you may be noticing patterns more clearly.

What's one system you use often that might be relying on these principles?

Next time you see a recommendation or autocorrect, ask yourself: what kind of learning is behind this? What is it assuming about me?

Chapter 3: A Short History of Smart Machines

AI didn't appear overnight. It's a product of decades of effort, trial, and imagination.

How does knowing that history change the way you think about the technology today?

Is there a past innovation, like the printing press or the internet, that this reminds you of?

Chapter 4: Your Digital Assistant

AI works quietly behind the scenes. That's part of its power, and part of the problem.

Have you ever relied on a "smart" system without realizing it was AI-powered?

Think about how often you trust those systems. Do you give them more power than you intended?

Chapter 5: At Work and in the World

AI in the workplace isn't just about robots; it's about decisions, efficiency, and sometimes, invisible oversight.

How would you feel if a part of your job were replaced or reviewed by an AI tool?

Would it feel like help? Or like surveillance? Would you want a say in how it was used?

Chapter 6: AI in Healthcare, Education, and Government

When AI enters the most human parts of life, the stakes rise. **Would you be comfortable with an AI helping decide your healthcare treatment or your child's education plan?**

What would give you peace of mind in that situation: transparency, choice, or a human back-up?

Chapter 7: Bias, Fairness, and the Limits of Algorithms

Bias in AI isn't hypothetical. It's already here, in quiet and dangerous ways.

Have you ever felt unfairly judged by a system, automated or otherwise?

Imagine what it would take to correct that bias. Who would you turn to for change?

Chapter 8: Can We Trust AI?

We trust systems every day, but trust should be earned, not assumed.

What kind of trust do you give to technology now? Is it different from the trust you place in people?

Think about a time when a system failed you. What did that experience teach you?

Chapter 9: Jobs, Automation, and the Future of Work

AI is transforming the workforce, not just replacing, but redefining roles.

How do you imagine your work changing in the next 5–10 years?

Are there parts of your job that could be automated? Which parts would you fight to keep human?

Chapter 10: AI and Creativity

If AI can write, paint, and compose, what does that mean for human expression?

What does creativity mean to *you*?

Would you feel proud of something made with AI's help? Or would it feel like cheating?

Chapter 11: Ethics, Laws, and What Comes Next

Governments are scrambling to regulate something that's already influencing millions.

Do you believe your local leaders understand how AI works?

If not, what kind of oversight or education would you want them to have?

Chapter 12: Becoming AI-Literate

AI literacy isn't about code. It's about questions. Awareness. Confidence.

What's one question about AI you couldn't have answered before reading this book, but can now?

And perhaps more importantly: **what new question will you keep asking?**

Note

As you close this book, know this: you don't need to become a technologist to understand AI. You need to stay aware, curious, and willing to ask better questions. The systems shaping our lives aren't beyond our reach; people, influenced by their choices, build them and remain open to change. AI may be complex, but your role in this moment is simple and powerful: **stay human, stay informed, and stay involved**. The future isn't something happening to you, it's something you get to help shape.

About the Author

Alex Hartman is a writer, educator, and lifelong explainer of complicated things. With a background in communication and digital literacy, Alex has spent over a decade helping individuals and organizations understand the technologies shaping our world, without relying on technical jargon or buzzwords.

Alex believes that artificial intelligence doesn't have to be intimidating and that everyone deserves a seat at the table when it comes to understanding the future. Through books, workshops, and online content, Alex focuses on making AI and emerging tech feel accessible, useful, and even a little bit empowering for non-experts.

When not writing, Alex enjoys long walks away from screens, good coffee, and conversations that start with, "Okay, but how does that work?"

The Book on AI for Everyday People is part of a broader mission: to bring clarity, calm, and curiosity into the way we talk about our increasingly digital world.

About the Publisher

Welcome to The Book On Publishing

At The Book On Publishing, we believe in rewriting the rules of learning. Whether you're chasing your next big idea, building a better life, or simply curious about what should have been taught in school, you've come to the right place.

We're a platform built for dreamers, doers, and lifelong learners, offering bold, practical books and tools that empower you to take charge of your journey. From real-world skills to mindset mastery, we publish the book on what matters.

No fluff. No lectures. Just what you need to know, delivered with clarity, purpose, and a spark of curiosity.

Start exploring. Start growing. Start writing your story.

Read more at https://thebookon.ca.

Acknowledgment of AI Assistance

Portions of this book were developed with the support of ChatGPT, an AI language model created by OpenAI. While every word has been carefully reviewed and refined by the author, ChatGPT served as a valuable tool for brainstorming, editing, and structuring ideas. Its assistance helped accelerate the creative process and bring clarity to complex topics.